Luigi Palma di Cesnola

A descriptive atlas of the Cesnola collection of Cypriote antiquities in the Metropolitan Museum of Art, New York

Luigi Palma di Cesnola

A descriptive atlas of the Cesnola collection of Cypriote antiquities in the Metropolitan Museum of Art, New York

ISBN/EAN: 9783741162817

Manufactured in Europe, USA, Canada, Australia, Japa

Cover: Foto ©Thomas Meinert / pixelio.de

Manufactured and distributed by brebook publishing software (www.brebook.com)

Luigi Palma di Cesnola

A descriptive atlas of the Cesnola collection of Cypriote antiquities

in the Metropolitan Museum of Art, New York

PLATE XXXI.

257. Gray terra cotta. Height, 4¼ inches. Found at Ormidia.

Rude figure of a warrior, with short, slightly tapering cylindrical body and conical base; somewhat in the style of No. 259. Top of helmet broken away; the remaining portion colored, and the nose and ear-pieces indicated, with red color. Right arm raised as in act to hurl a weapon, but the hand is too much defaced to show any trace of a weapon there. On the left arm, a round shield, or targe, held almost horizontal, ornamented with brown lines crossing at the centre, which alternate with shorter red rays. Eyes and beard indicated in brown; bands of brown and red on the right arm; brown band at junction of body with base; on the base, alternate stripes of brown and red from top to bottom. Surface somewhat defaced. Broken across the body, and reset.

258. Gray terra cotta. Height, 6⅝ inches. Found at Dali.

Rude figure of a warrior; cylindro-conical body not turned on a wheel; helmet, pointed conical, with ear-pieces. On the left arm, ancient form of elliptical shield, with indentations on opposite sides of the narrower part; right arm upraised, but broken away at a little distance from the shoulder. Eyes, mouth and beard indicated in brown color; helmet-tip colored red; traces of red and brown color throughout on the body and shield.

259. Gray terra cotta. Height, 5¾ inches. Found at Ormidia.

Rude figure of a warrior, with cylindrical body and conical base turned on a wheel; right arm raised in act to hurl a spear—the latter now broken away; on the left arm a round shield or targe. Helmet with long point, or crest, curved and hanging forward; the face upturned so that the chin and long beard extend almost horizontally. Helmet-crest and nose-piece, and the shoulder-straps, colored red; shield colored in alternate sectors of red and brown, separated by brown stripes; eyes, eyebrows and beard indicated in brown; right arm with bands of red and brown; shoulder-straps red; conical base decorated with a band of brown at top and

PLATE XXXI. CONTINUED.

bottom, between which are triangular spaces of alternate red and brown. Hole pierced in the base, as if for suspension. Broken across the base, and reset.

260. Gray terra cotta. Height, 3¼ inches. Found at Temple of Apollo Hylates, near Curium.

Fragment of a figure of somewhat the same description as No. 261. Helmet with central top, curving forwards, and two minor pieces diverging upwards from the sides—all three broken away at the tips. Cheek and chin protectors very prominent, in the shape of flaps from the back of the head, fastened at the chin. Shield round, with female head, probably of Medusa, in relief, surrounded by a braided wreath, which passes behind the front hair at top, and is tied in a bow under the chin. Originally colored red throughout, but apparently in varying shades, or with tones of another color. A large portion of the right arm, and small portions of the edge of the shield, broken away ; all the lower parts of the figure, from the waist down, wanting.

261. Red terra cotta. Height, 8 inches. Found at Temple of Apollo Hylates, near Curium.

Figure of bearded mounted warrior, with crested helmet ; round shield on the left arm ; right arm nearly all broken away, but the stump indicating that it was uplifted to hurl a weapon. Helmet with forward-bending crest and front portion all in one piece ; neck-protector behind, and chin-strap in front. Shield showing the head of Medusa in relief, inside a rim whose surface is a little lower.

Horse ; the head bold, but rude, with exaggerated forelock. The whole figure formerly colored red. Traces of trappings on the horse, and of decoration throughout, in some ill-defined color—either light brown or light red. Right arm and leg of horseman, with right hind leg, part of tail and of left foreleg, broken away.

262. Red terra cotta ; exterior now gray. Height, 4¼ inches. Found at Dali.

Fragment of figure of a warrior, with cylindrical body. Top of helmet, or cap, curving forward, but its tip broken away ; heavy rolled brim about the front and sides. On the left arm a large round shield made of a disk raised a little higher than the broad rim. Features and beard moulded. Traces of red color here and there. Right arm, lower part of the body, and the greater part of the shield, broken away,

PLATE XXXI. CONTINUED.

263. Gray terra cotta. Height, 4⅜ inches. Found at Ormidia.

Very rude figure of a warrior, with cylindrical body and slightly flaring base, not made on a wheel. Round shield or targe on left arm; the right arm uplifted as if to throw a weapon. Helmet rudely indicated by two strips of clay colored red. Eyes and beard indicated in brown color; ears large, ornamented with brown. Shield with central boss, about which are concentric rings of color, alternately red and brown. Part of helmet, end of right arm, and the upper portion of the shield broken away.

264. Red terra cotta. Height, 5 inches. Found at the temple of Apollo Hylates, near Curium.

Fragment of the figure of an archer. Head turned to the left; stumps of arms in position as if in the act of discharging an arrow, the right hand having been held up nearly to the right ear, as appears from fractures and material on the neck at that spot. Quiver slung over the shoulder, with a rude indication of the ends of arrows in it. Quiver straps indicated in front by incised lines; fringed girdle also similarly indicated. Heavy collar and necklace. Head-dress a flat-topped, close-fitting, low cap. Beard marked with incised lines. Surface and features defaced. Nose broken away, as are also both arms and all the body from a little below the girdle.

265. Red terra cotta, exterior now reddish brown. Height, 4 inches. Found at Dali.

Fragment of figure in much the same style as No. 260, but no side-elevations on the helmet, nor division in the cheek- and chin-pieces. Shield oval, with flat depressed rim, with the remains of some device which now looks like a ram's head and horns, or else a headless flying bird. Most of the right arm, part of the top of the helmet, a great portion of the shield, and all of the body below the waist, broken away.

Plate XXXII.

266. Red terra cotta. Height, 6⅛ inches. Found at Soli.

Draped female figure, perhaps Demeter, in early Greek style. On the head a wreath of roses, beneath which appears the hair in twisted strands about the forehead; a peplos falling from behind. Other dress, a chiton and chlamys. Right hand holds (apparently) a vase (or fruit?) to the breast; the other hand, pendent, holds a part of the dress. Folds and texture of dress finely moulded. Chiton gathered to a circular brooch at the neck. Very large feet. Somewhat defaced in spots.

267. Red terra cotta, exterior now gray. Height, 6¾ inches. Found at the same place as the last.

Draped seated female figure, apparently Demeter. From the low head-dress falls a peplos. Shape and dress evidently conventional, somewhat resembling the Roman *Terminus*. Arms either absent or concealed in the drapery.

268. Red terra cotta, exterior now gray. Height, 6½ inches. Found at Amathus.

Draped Aphrodite; right arm folded in peplos to the wrist and held to left breast; left arm pendent, holding the peplos from within. Head-dress, a hooded peplos, falling behind, but enveloping arms and shoulders as above. Large ear-rings; throat-necklace, beads with brooch; other necklace, a chain with medallion. Heavy wristlets on right arm. Dress, a diploïs and chiton, the latter falling in front in broad plaits.

269. Buff terra cotta. Height, 6¼ inches. Found at Kiti.

Draped female figure; chiton held in circular clasps at each shoulder; the diploïs may be one garment with the chiton. Both hands gone, leaving some doubt respecting the character of the figure. Head-dress pointed, with peplos flowing from its apex. Hair in thick tresses over the forehead.

270. Red terra cotta, exterior now gray. Height, 3½ inches. Found at Cythrea.

PLATE XXXII. CONTINUED.

Draped Aphrodite; head-dress pointed, with peplos; much like Plate XXIII., Nos. 182, 184, and one of a series which includes also Nos. 272, 273, though the individuals present many minor variations. Chiton reaching to the feet; chlamys held together above, but open below; hands held to the breast on either side. Throat-necklace with large clasp.

271. Gray terra cotta. Height, 8¼ inches. Found at Cythrea.

Draped female figure, perhaps Aphrodite; right hand held to breast; left arm pendent; oval flaring head-dress (hat), beneath which appear heavy braids and tresses. Surface much defaced, so that the dress cannot be accurately described, but certainly a long thin chiton displaying the contour of the form, with probably a peplos also.

272. Gray terra cotta. Height, 6⅛ inches. Found at Dali.

Draped female figure, perhaps Aphrodite; much like Plate XXIII., Nos. 182, 184. High pointed head-dress, with peplos falling behind to feet. Hair in thick twisted strands. Traces of a throat-necklace. Dress, a chiton and diploïs, of which the hem of each is visible at the neck, showing them to be separate garments. The hands, pendent, grasp the chiton and peplos. Broken a little along the edges.

273. Gray terra cotta. Height, 6¼ inches. Found at Cythrea.

Aphrodite, in same series as the last, which it closely resembles; but the pointed head-dress is curved a little upward; the peplos is more ample; a *tænia* below the breasts is held by a circular clasp between them; two necklaces; one at the throat, with large beads and brooch; the other, with pendants, hides the hem of a chiton.

PLATE XXXIII.

274. Red terra cotta, exterior now gray. Height, 6¼ inches. Found at Soli, or Cythrea.

Draped female dancer. This is one of a numerous series, some of which, like this one, have a pointed head-dress, but others a flat-topped one. Peplos hanging from the head-dress, down the back and over the arms; hair in a roll over the forehead, with tresses hanging to the shoulders; throat-necklace with beads and pendant; a second necklace with pendant between the breasts; chiton; himation over the left shoulder, winding about the body below the breasts ike a large sinus (or *kolpos*). The figure seems to be a dancer in the character of Aphrodite. The series belonged to groups of five, dancing in a circle, with a central figure. Right arm and left hand broken away.

275. Red terra cotta. Height, 6⅝ inches. Found at Soli, or Cythrea. One of the same series as the last, but at rest.

Figure like the Aphrodite of Plate XXXII., Nos. 272, 273; pointed head-dress; peplos reaching to the feet; throat-necklace with circular brooch; two lower necklaces, one with beads, the other with pendants; tænia holding diploïs by a circular clasp between the breasts; chiton reaching to the feet. Extension from the shoulders, like the arms of No. 274, as if the arms of a figure behind, dancing in a ring about Aphrodite. One of these extensions partly broken away.

276. Red terra cotta, exterior now creamy gray. Height, 6¼ inches. Found at same place as last.

Figure of the class mentioned at No. 274, and of nearly the same description. Both arms broken away.

277. Red terra cotta, exterior now creamy gray. Height, 4⅝ inches. Found at same place as last.

Upper portion of figure of same series and description as No. 274; but round flat-topped

PLATE XXXIII. CONTINUED.

head-dress. Locks of hair falling down over the shoulders in front; chiton with short sleeves; throat-necklace with pendant; bracelet on right wrist. Left arm broken away.

278. Red terra cotta, exterior now creamy gray. Height, 4½ inches. Found at same place as last.

Upper portion of figure in same series; low cylindrical head-dress; ample peplos, falling over projections from the shoulders, as well as behind. Hair beneath the peplos in twisted rolls about the forehead. Throat-necklace with pendant; diploïs; chiton; right arm held across the waist; bracelet on the wrist; wreath in the hand; left arm invisible. Extensions from the shoulders partly broken away.

279. Gray terra cotta. Height, 3⅛ inches. Found at Lapithus.

Three rude male figures, with extended arms and pointed cap, dancing about a fourth, who is playing the double flute. Traces of brown color here and there. One arm of each of the dancing figures, and both arms of one of them, with one pipe of the double flute, partly broken away.

280. Gray terra cotta. Height, 5⅝ inches. Found at Soli, or Cythrea.

Figure of same character and series as No. 275; but point of head-dress curving higher; a himation winding about the body from the left shoulder below the breasts, like No. 274. Right arm as in No. 275; left arm holds some object, apparently a vase, upright between the breasts. A thick fold of the peplos falls over the left forearm, near the elbow, almost down to the feet. One of the extensions from the shoulder partly broken away. Broken across and reset.

281. Red terra cotta, exterior now gray. Height, 6½ inches. Found at Soli, or Cythrea.

Figure of nearly the same description as No. 275; but head-dress nearer the perpendicular. Extensions from the shoulders broken off close.

PLATE XXXIV.

——— — ———

282. Red terra cotta, exterior now gray. Height, 6¼ inches. Found at Cythrea.

Female lyre-player; holding lyre in left arm; the right, with bracelet on wrist, folded across the body, holding a plectrum. Low turreted head-dress, with ample peplos falling behind; hair in a roll over the forehead, with side tresses; dress, a chiton and diploïs. Broken across, and reset.

283. Buff terra cotta. Height, 5½ inches. Found at Kiti (Salines near Larnaca).

Female lyre-player, seated. Lyre (cithara) held in left arm, and played with the right hand. Hair combed back from the forehead, and gathered in a knot behind. Head-dress, a tiara; peplos revealing the coiffure beneath, and falling down the back. Chiton clasped over the right shoulder, leaving the right arm bare. Rear portion somewhat broken away.

284. Red terra cotta, exterior now gray. Height, 6⅝ inches. Found at Cythrea.

Female lyre-player; pointed head-dress; peplos; diploïs; chiton; bracelet on wrist; hand holding peplos, as in No. 282.

285. Red terra cotta, exterior now gray. Height, 9¾ inches. Found at Lapithus.

Almost identical with, and probably made in the same mould, as Plate XXV., No. 202, and there described. Here shown in a slightly different position, as well as in a different grouping, to which it equally well belongs.

286. Red terra cotta, exterior now gray. Height, 12¼ inches. Found at Cythrea.

Female lyre-player; turreted head-dress, from which falls a peplos behind. Chiton clasped over right shoulder; apparently a himation comes over the left shoulder, under the cithara, and winds about the body. Apparently the neck-band of both chiton and diploïs are visible, as separate garments; but the latter may be the remnants of a necklace. Formerly colored light brown. Surface much defaced. Style late Greek. Broken, and reset.

287. Gray terra cotta. Height, 10⅜ inches. Found at Lapithus.

PLATE XXXIV. CONTINUED.

Female lyre-player. Head-dress a tiara of several rows of pearls ; peplos falling behind, but gathered up over and falling within the right arm ; a few folds of it also falling from the left arm below the lyre. Hair in a roll over the forehead, falling also in curls that spread over the shoulders in front. Chiton and diploïs ; from beneath the latter fall the two ends of a wide girdle (zonion) with broad transverse embroidered bands and long terminal fringes.

Throat-necklace clasped by a heavy brooch with pendants ; a second necklace with pendants ; the third, a chain supporting a wreathed bar and medallion. Bracelet on right wrist, which might easily be confounded with the folds of the peplos. Traces of red color throughout.

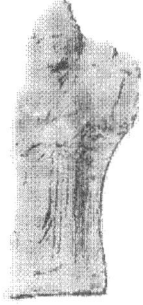

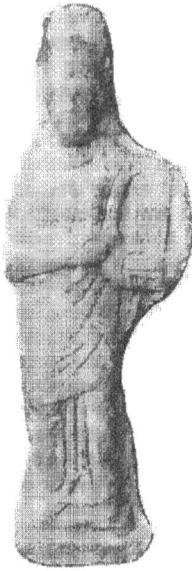

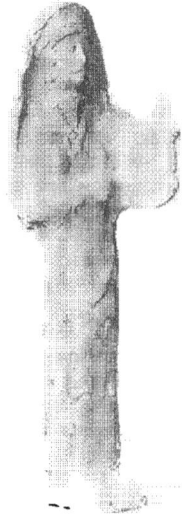

PLATE XXXV.

All the objects on this plate found at the temple of Apollo Hylates, near Curium.

288. Red terra cotta, exterior now gray. Height, 7¾ inches.

Male head, bearded; with appearance of early manhood. Hair and beard wrought in curls by incised marks. Hair confined by a band reaching entirely about the head, but wider over the forehead than elsewhere. Features rather heavily moulded, especially the eyebrows and eyelids. Hair and beard colored dark brown on the raised portions. Eyebrows and eyelashes colored dark brown; traces of red color here and there. End of nose, top of head, and right ear somewhat abraded.

289. Red terra cotta, exterior now grayish. Height, 9 inches.

Male head, beardless; hair indicated by incised lines; fragments of a wreath of leaves still remaining about the sides of the head and brow. Lower line of the nose elevated at an angle of about 50°, showing the nostrils prominently. Ears small, placed rather high. The whole once colored with red color, of which traces remain. Slight abrasions on tip of nose and on the ears.

290. Red terra cotta, exterior now gray. Height, 18 inches.

Rather rude figure of a person clothed in a *stole*, or robe reaching to the feet, the body turned on a wheel and afterwards modified in shape. Head beardless; face slightly uplifted; nose and eyes very prominent; the ends of the hair are seen on the forehead beneath the shallow, close-fitting skull-cap with very narrow brim. Stump of right arm uplifted, but nearly all of that arm, as well as all of the left, broken away. The left foot projects nearly half its length from the general line of the bottom of the robe, but is still covered by the latter; the right foot is broken away. On each side, from the arms down and also in the middle in front, decorative bands of red color, formed of heavy vertical lines joined by oblique transverse strokes, extend to the foot of the robe. Head broken off, and reset. Portions broken away on rim of cap and bottom of robe, with slight chips here and there. Surface somewhat incrusted and marred.

PLATE XXXV. CONTINUED.

291.   Gray terra cotta.   Height, 9 inches.

Upper part of figure clad in a tunic open at the right shoulder; over the head a peplos wrinkled or ruffled a little in front, descending behind, and spreading so as to cover half of each shoulder.   Beneath the peplos the hair appears over the forehead in small bandeaux.   Features sharply marked, especially the eyes and eyelids; the pupils marked with a punctured dot. Traces of a uniform color, probably reddish brown, over the whole.   Slightly chipped in spots.

292.   Red terra cotta, exterior mostly grayish from incrustation.   Height, 7½ inches.

Fragmentary and rather rude head and bust, probably of Aphrodite; with hoop-like crown about the head; the whole once colored red.   Large portion on the left side of the head broken, and reset; part of nose and of the left lower jaw and of both ears, chipped away; both arms broken away at the shoulder.

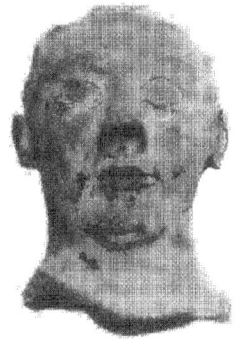

PLATE XXXVI.

All found on the site of the Temple of Apollo Hylates, near Curium.

293. Red terra cotta, exterior now whitish gray. Height, 11½ inches.

Seated, naked figure, probably devotee at a shrine or temple. Figure seated on the ground, the left leg bent so that the foot is directly in front of the body ; the right leg so that the foot is planted on the ground. Left arm stretched straight to the ground, supporting the weight of the shoulder by the palm of the hand. Right hand open, palm upwards, resting on the knee. Head beardless, with pointed chin, strongly expressed features, large eyes and ears, hair thin and wrought by incised lines. Remnants of a string of amulets, that doubtless once hung over the left shoulder and under the right arm, are present on the right side, but not visible in this plate. Whole figure originally colored red. Head and neck broken off, and replaced ; ends of fingers of right hand broken away ; and slight abrasions in other spots.

294. Red terra cotta, exterior now mostly gray. Height, 9½ inches.

Beardless figure, seated on the ground, in nearly the same posture as the last ; but clad in a short tunic which covers the shoulders, the arms nearly to the elbows, and the body down to the knees. Feet and legs also seem to be clad in close-fitting boots or stockings. Hair like a close-fitting skull-cap.

The left hand grasps something on the ground ; but abrasions make its character doubtful.

Head broken off, and reset, the fractures not admitting exact joining. Right arm broken away at the elbow.

295. Red terra cotta, exterior now brownish gray mottled with white. Height, 6½ inches.

Rude figure seated on the ground ; belt or string of amulets hanging from both shoulders down over the chest and nearly to the waist. Arms as if akimbo, but hands resting on the legs ; which latter are broken away. Hair wrought in horizontal incised strokes. Broken in several places, and reset.

PLATE XXXVI. CONTINUED.

296. Height, 9¼ inches.

Beardless draped figure, much in the same posture as No. 294, and draped nearly in the same way, but the sleeveless tunic has armholes at the shoulder. A string of amulets hangs over the right shoulder and under the left arm. Head broken off, and reset. Both arms broken away near the shoulder. Abrasions in various places.

297. Red terra cotta, exterior now somewhat grayish. Height, 13⅝ inches.

Beardless figure seated on the ground, of nearly the same description, posture, and clothing as No. 294; but the hair indicated by incised markings; each hand upon its corresponding knee; a string of amulets over the left shoulder and under the right arm; decided folds or wrinkles visible in the tunic about the left arm; bracelet on left wrist, and traces of an armlet on the right forearm below the elbow. Head broken off, and replaced, but the joining could not be made close; right hand, with parts of fingers of left hand, broken away. Slight abrasions in spots.

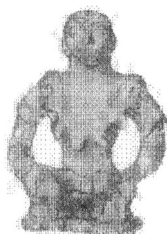

THE.

PLATE XXXVII.

298. Red terra cotta. Height, 6¼ inches. Found at Dali.

Beardless head, with pointed nose and chin; ornamental band or crown about the head, which seems to consist of several oblong pieces, each adorned with a disk in the centre and curved outward a little at the ends—probably imitating a metallic crown. On the sides of the neck, and outwards towards the shoulders, fall either bunches of hair or the lappets of a head-dress—it is uncertain which by reason of the fragmentary state of the figure. On the neck, to the left, is the fragment of a chain necklace. The whole figure originally colored red, with perhaps a different shade or another color on the eyebrows and mouth. The crown a little abraded in spots.

299. Gray terra cotta. Height, 7¼ inches. Found at Dali.

Beardless head (female?), with pointed nose and sharp-edged square chin. Hair wrought, by incised marks, in curls all over the head, confined by a band on which are rosettes at regular intervals. Ear-rings of heavy spirals. Right ear-ring mostly broken away, with part of the ear.

300. The same figure as No. 293, Plate XXXVI., and there described. (When Plate XXXVI. was first made, the head had not been found, and that figure was drawn headless. After the head was found, and the whole figure inserted on this plate, a delay in printing allowed the colored plate, XXXVI., to be corrected also.)

301, 302. Front view (nearly) and side view of the same object. Red terra cotta. Extreme height, 11¼ inches; extreme length and breadth of the base, 12 x 7 inches. Found in a tomb in the vicinity of Ktima, or Neo Paphos.

Draped figure on a sloping couch or divan, the elbows resting on the low arms of the latter; clad in a tunic that seems drawn together in front over the breast, showing the folds and wrinkles by incised marks. The lower part of the figure is covered by a loose mantle or coverlet, whose irregular folds are wrought with some freedom in relief. From beneath this coverlet the feet peep out slightly, showing the toes of the right foot, the sole and great toe of the left. Head and hands broken away; slight abrasions on the bottom corners of the couch. Whole figure broken across the lap, and reset.

PLATE XXXVIII.

303. Red terra cotta, exterior now grayish salmon color. Height, 6⅜ inches. Found at Cythrea, or Soli.

Female figure, with low turret head-dress; traces of a peplos; chiton reaching to the feet, displaying the form of the right leg; diploïs; throat necklace with large pendant; right arm bare from the elbow, held across the waist; bracelet on the wrist, and hand holding a wreath; left arm invisible; hair in twisted rolls about the forehead. One of the series remarked at Plate XXXIII., No. 274; and almost or quite identical with the fragmentary figure No. 278. Extensions from the shoulder, like the arms of certain other figures of the series, suggest the arms of a dancer in the rear; but both are broken away.

304. Gray terra cotta. Height, 7₁⁵₆ inches. Found at same place as the last.

One of the same series just noted, and of essentially the same description as the fragment No. 277. Right arm broken away.

305. Red terra cotta, exterior now gray. Height, 6₁⁸₈ inches. Found at the same place as the last.

One of the same series; very high pointed head-dress, and very ample peplos; otherwise as No. 274, Plate XXXIII. Both arms broken away.

306. Gray terra cotta. Height, 6₁⁷₂ inches. Found at same place as the last.

Figure of same style as the last, but smaller head-dress and less ample peplos. Left arm broken away.

307. Gray terra cotta. Height, 7⅜ inches. Found at Cythrea, or Soli.

Youthful figure, probably Græco-Roman. Head-dress a turban, with clasp in front, and ample peplos behind, held by the arms; the hands apparently clasped behind the head. Hair parted in the middle and hanging in tresses upon the shoulders in front. Dress, drawers, and a tunic belted at the middle by a narrow girdle with long hanging ends; a sheathed dagger on left

PLATE XXXVIII. CONTINUED.

thigh. Figure apparently a dancing girl in character costume; the head inclined to the right, the body curved in the same direction; the right leg raised. Upper part of the top, including both hands, broken away, as is also a large portion of the lower parts, including the feet.

308. Gray terra cotta. Height, 5¼ inches. Found at same place as last.

One of the same series remarked at Plate XXXIII., No. 274. Head-dress pointed, the point extending backwards almost horizontally, and not visible in the plate; supporting a peplos which spreads out to the elbows on the extended arms, and then falls down to the feet. That part of the head-dress visible below the peplos is adorned with points and small knobs, like a crown; beneath it the hair is adorned with a string of pearls or large beads, extending down to meet the throat-necklace. Thin long locks of hair descend over the shoulders in front down beside the breast. Throat-necklace of large beads, with clasp in front; another necklace just above the breasts. Clad in a sleeved chiton, covering but displaying the breasts, and falling in vertical folds to the feet.

309. Red terra cotta, exterior now gray. Height, 6¼ inches. Found at Lapithus.

Figure of design similar to the last, but much ruder in execution. Head-dress a hooded peplos, quite ample. Arms rudimentary. Over the left shoulder seems to come a frilled or figured sash, passing below the breasts and under the right arm; but it may be the border of an enwrapping himation.

310. Red terra cotta, exterior now gray. Height, 7½ inches. Found at Cythrea.

One of the series mentioned above, and of like description with No. 304. Part of both arms and extensions broken away.

311. Gray terra cotta. Height, 6¼ inches. Found at same place as the last.

One of the same series, and closely resembling Nos. 275, 280, 281, of Plate XXXIII. Pointed head-dress, broken at the tip; peplos; diploïs; chiton; three necklaces; tænia holding diploïs with a circular clasp between the breasts. Extensions from the shoulders and part of left foot broken away.

305

307
308

THE GE' - ...A
liohac.

PLATE XXXIX.

312. Red terra cotta, exterior now gray. Height, 5¾ inches. Found at Kiti, at the Salines near Larnaca, in the vicinity of the ruins of the temple of Artemis Paralia.

Figure of an attendant of Demeter, with *cista*. Hair gathered with a jewel ornament at top of forehead, whence it is combed back in curving bands to the back of the head; whence a peplos falls down behind. Dress, a chiton clasped at the shoulders, with diploïs confined by a *zonion* just below the breasts. The peplos covers the left arm behind the *cista*, but hangs in the rear of, and touching, the naked, pendent right arm.

313. Red terra cotta, exterior now mostly blackish gray. Height, 6 inches. Found in same locality with the last.

Female figure, in good Greek style; hair dressed high above the forehead; peplos falling from the back of the head; large ear-rings; necklace. Dress, a chiton, and a himation beneath which the arms are folded, as if carrying something in the left arm. Left knee bent slightly forward. Has the general air of a woman walking in a religious procession.

314. Red terra cotta. Height, 5¼ inches. Found near the last.

Figure of youth, wearing a low petasus, a chiton, and chlamys; the latter covering the arms. Right arm pendent; the left bent from the elbow across the breast. Of good Greek style.

315. Red terra cotta, exterior now gray. Height, 4¼ inches. Found near the last.

Figure of attendant of Demeter, with *cista* in left hand; right arm bent at the elbow, and the hand holding the himation together, over the breast. Other dress, a chiton, close-fitting, revealing the body, except where it falls in perpendicular folds. Hair dressed high above the forehead, in the style of certain classes of attendants of the goddess.

316. Red terra cotta. Height, 10¼ inches. Found at same place with the last.

Female figure, with hair dressed in two high masses behind, from which falls a peplos; dress, a chiton and diploïs belted by a cord-like *zonion* tied in front. The left hand holds a large

PLATE XXXIX. CONTINUED.

cista; the right hand (the arm pendent) grasps the folds of the peplos. Doubtless an attendant of Demeter in a religious procession.

317. Slaty-red terra cotta. Height, 5¼ inches. Found at same place as the last.

Female figure, of good Greek style; hair in rolls at the sides of the face, with a knot behind; perpendicular crescent-shaped crown a little behind the forehead; himation wrapped closely about the neck and body, but held up in front somewhat by the two arms, which are bent forward at the elbows, but concealed beneath the garment; beneath the himation appear the folds of the chiton, reaching to the feet.

318. Red terra cotta. Height, 5¾ inches. Found at same place as last.

Sitting figure of Demeter; head covered by a hooded peplos, beneath which, over the forehead, appears the hair adorned with a wreath of roses. A himation covers the body from shoulders to below the knees, held up from within by the left arm, and concealing the right arm, that rests on the right knee. Below the himation is seen the chiton, falling in folds to the feet. Throne ornaments at the right shoulder partly, those at the left wholly, broken away. Cracked, and reset.

THE GG...
LI Jin

PLATE XL.

319. Red terra cotta. Height, 6 inches. Found at the temple of Artemis Paralia, at the Salines near Larnaca.

Female figure, in good Greek style; of nearly the same description as No. 327, Plate XLI., but the right arm pendent, the left bent at the elbow, with the hand upon the hip. The himation covers the body closely, revealing the covered bust and arms; it is gathered up a little by both hands within. Below it, a chiton falls to the feet and conceals them. Head broken away.

320. Red terra cotta. Height, 5½ inches. Found at same place as last.

Female figure, of good but rather late Greek style, standing on a base; the head broken away. Chiton reaching and concealing the feet; himation wrapped about the neck, thrown back over the left shoulder, whence it hangs in thick folds; the left arm, wrapped in one of its folds, but outside of the main portion, holds it at the neck. The right arm, pendent, grasps the himation from within, making a wrinkled place in the portion that falls in front.

321. Unbaked, or very slightly baked, clay; the same as that of the red terra cotta. Height, 6¾ inches. Found at same place as last.

Fragment of a lyre-player, apparently; the head, and all the left side down to the hips, broken away. From the head descended a most ample peplos. Clothing, a chiton, belted at the waist with a wide ornamental girdle, clasped at the shoulder, leaving the arm bare, and falling to the ankles. Feet large, in shoes fastened with tasseled laces. Left arm (the other is gone) bent at the elbow, wearing a heavy armlet, and holding a plectrum in the hand.

322. Red terra cotta. Height, 4 inches. Found at same place as last.

Female figure, of rather late Greek style; draped in a himation somewhat hooded over the head and reaching to the feet. Below it is seen a chiton. The left arm, beneath the garment, rests on the hip; the right arm, outside the garment, is very large, deformed, and

PLATE XL. CONTINUED.

stumpy; held off from the shoulder, and hung unnaturally. Right hand broken away; chips broken off the left side of the face, and at the base.

323. Red terra cotta. Height, 6 inches. Found at same place as the last.

Female figure, of good Greek style. Hair with three horizontal rolls, one above another, and all above the parted tresses about the forehead. From the top a peplos falls behind. Earrings; necklace; chiton falling to the feet; himation over the shoulders, gathered up over and within the arms; arms folded across the chest. Large piece broken away near the base. Surface much defaced and incrusted.

324. Red terra cotta. Height, 4⅜ inches. Found at same place as the last.

Female figure, much worn and defaced. A large himation, hooded over the head, and falling on the shoulders, hangs away from the body on the left, but is drawn about the body from the right, covering the left arm and the breast, whence it falls nearly to the feet. At the neck is seen the chiton, covering the right breast; and also below the himation, concealing the feet. The drapery reveals the form and limbs. Left arm akimbo; the right arm broken away.

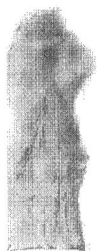

PLATE XLI.

All found at the temple of Artemis Paralia, at Kiti, at the Salines near Larnaca.

325. Red terra cotta. Height, 5 inches.

Fragment of female figure, of good Greek style; seated; enveloped in a himation which is held together at the neck by the concealed left hand, and upon the lap by the covered right hand. Chiton visible from the knee down. Head inclined to the right; hair with high wavy ridges along the part in the middle, the rest combed back and massed behind the neck. Stocking on the foot. Colored red. Left side and limb, from the waist down, broken away.

326. Buff terra cotta. Height, 5¼ inches.

Fragment of male figure with long beard, perhaps Bacchus; the head, right arm, and feet broken away. A himation wrapped about the left shoulder and under the bare right arm, and again over the left arm. Below this himation is a short garment, probably an *exomis*. The left hand carries a small vase. Figure once colored red.

327. Buff terra cotta. Height, 6¼ inches.

Female figure, of good Greek style; the head broken away. A himation wrapped about the shoulders, and held to the mouth by the concealed right hand; covering also the pendent left arm. Beneath this wrapping a portion of the himation falls a little below the knee. Under the himation a chiton, reaching to the feet and concealing them.

328. Red terra cotta, exterior now gray and white. Height, 7¾ inches.

Female figure, of good Greek style; hair dressed in banded puffs, so as to resemble a melon, and gathered in a knot behind; small ear-rings; arms folded across the chest, but left forearm broken away. Dress, a chiton and diploïs, the two apparently one garment, clasped at the shoulders and leaving the arms bare. The chiton trails on the ground, concealing the feet. Broken, and reset.

329. Red terra cotta, exterior now gray. Height, 8¾ inches.

PLATE XLI. CONTINUED.

Woman and child ; the latter naked, held in her left hand in a sitting posture, almost upon her shoulder; his hand playing with her hair. The woman's head is bent a little forward, hair apparently as that of No. 328, but covered nearly to the forehead by a peplos which falls bunched behind. Small ear-rings. A himation over the left shoulder winds about the body under the left arm, in which both its edges are gathered up under the child, whence its folds fall downward. Below the himation is a chiton that apparently reached to the feet, but its lower portion is broken away. Traces of the chiton are seen also above the himation on the right side ; its clasp, which belongs upon the right shoulder, having fallen down a little way on the right arm, leaving the right breast and shoulder bare. The right arm is likewise bare, except where the chiton clasps about it; having a bracelet at the wrist.

330. Grayish-red terra cotta, exterior now white and brown. Height, 7½ inches.

Female figure, of good Greek style, standing on a base. Head inclined a little forward, and to the left, with an expression of inquiry. Hair in thick waves, uniting in a knot at the back, and having also a thick high frisure at the top. Body almost entirely wrapped in a himation, which reveals the pendent left arm, the breasts, the advanced left leg, and the right arm. The last, bent at the elbow, holds up a part of the robe in a kind of sinus that reaches to the left hand. Below the gracefully falling himation appears the chiton on the right—the himation reaching to the left foot, which, covered with a shoe, protrudes a little from the garment.

Nos. 327, 328, 330, No. 319 on Plate XL., and several others of the Cesnola terra cotta statuettes, are of the same class and make as those since found at Tanagra, Myrina, and other localities.

325

328

326

TK.

PLATE XLII.

All found at the temple of Apollo Hylates, near Curium.

331. Red terra cotta, exterior now gray. Height, 2¼ inches.

Winged Eros, recumbent on his left side, drinking from a *phiale*, which is held in his right hand. Part of the right wing, and all of the left arm, as well as portions of the base and background, broken away. Of good Greek style.

332. Gray terra cotta. Height, 2¾ inches.

Fragment, in high relief, of a youthful male figure embracing a winged female figure. He stands on the right, his left arm around her from behind, with the hand over her left arm (which is stretched away—as if reluctant), as if pushing off the drapery which still covers it; his right arm in front, its hand grasps her left shoulder. The drapery over her left arm, grasped by her hand, and falls in folds; similar falling drapery visible on her right. Heads of both figures broken away, as also portions of the wings of both, and from the background of the relief. Surface somewhat defaced. Probably Cupid embracing Psyche, or else Zephyr about to carry Psyche to the palace of Cupid.

333. Buff terra cotta. Height, 5⅞ inches.

Boy, standing; nude, except a himation partly covering the head behind, falling down the back, and thrown over the left arm. Left arm bent at the elbow, the hand holding an object like a ball or fruit; right arm pendent, holding in the hand an object that is indistinct, but in all probability a round hat, or pilleus. Of good Greek style.

334. Red terra cotta, exterior now gray. Height, 3½ inches.

Figure of nude boy, of almost exactly the same description as the last, except that the himation is simply hung over the left arm; the right arm bent at the elbow, the hand holding against the chest a pilleus that is folded, instead of being rounded out as in No. 333. Head broken away. Broken, and reset.

PLATE XLII. CONTINUED.

335. Red terra cotta, exterior now reddish brown. Height, 4½ inches.

Figure of boy, of same general description as Nos. 333, 334, but the right hand resting on the hip; himation over the head and about the neck, thrown off from the right shoulder behind, and hanging in front across to the left side, covering the left arm and the object held in the left hand. Surface much defaced, and front part of head broken away.

336. Buff terra cotta. Height, 3⅝ inches.

Figure of a boy astride a kneeling-recumbent goat; his right hand raising a *phiale* towards his mouth. Over his shoulders a short himation, or chlamys. Probably the infant Bacchus. One end of the base, left foot of the child, and the head of the goat, broken away. Broken, and reset.

337. Red terra cotta. Height, 2 inches.

Figure of nude child, seated on a bench or stool. Right hand rests on the edge of the seat; the left hand, resting on the left knee, holds a bird. Of good Greek style. Head broken away.

338. Red terra cotta. Height, 2½ inches.

High relief, with base, representing two winged children—either two Erotes or Cupid and Psyche—walking, each with an arm over the other's shoulder. The one on the left has a mantle hanging in thick folds over the left arm, which is bent at the elbow. One wing of each figure is visible. The background is somewhat damaged, and portion of the figure, with the robe, somewhat defaced. Surface somewhat worn.

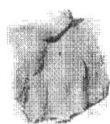

THE

PLATE XLIII.

339. Red terra cotta. Height, 6¾ inches. Found at Soli.

Female figure of rather archaic style, standing on a pedestal, and holding a mesomphalic *phiale* in the right hand. Head with curly hair on the forehead, but covered by a small peplos, which hangs in a bunch behind the neck. Dress a close-fitting chiton, reaching to the (very large) feet, but not covering them; its hem seen about the neck, and its folds between the shins below the himation. The latter is thrown over the left shoulder and arm, drawn about the body under the right arm, and falls a little below the knees. Both arms pendent.

340. Red terra cotta. Height, 6¼ inches. Found at Cythrea.

Relief of naked archer, with Phrygian cap; bow in left hand; an arrow (now broken off) in the right hand; quiver-strap over right shoulder; a loose himation hanging over left arm. Left leg advanced, bent at the knee; figure advancing to the left, as if ascending a hill.

341. Gray terra cotta. Height, 6¾ inches. Found at Cythrea.

Draped female, standing with left leg thrown over the right hand at the hip; left hand on a supporting pillar. Hair hanging down upon the shoulders in front; throat necklace with clasp; dress a short-sleeved chiton, and a himation thrown about the body from the left shoulder, covering the left arm. Right arm bare.

342. Red terra cotta. Height, 6¼ inches. Found at Cythrea.

Figure of a youth, wearing a petosus; short-sleeved, belted chiton reaching to the knees; chlamys thrown about the left shoulder, and clasped over the right, grasped by a hand on either side. Right arm mostly uncovered; left arm and hand covered by the chlamys. Dagger in the belt on the right side.

343. Red terra cotta. Height, 7¾ inches. Found at Cythrea.

Female figure, on a base; hair in a braid about the forehead; braided wreath about the head; very long neck; head turned a little to the right; the right hand on a supporting pillar;

PLATE XLIII. CONTINUED.

left hand held behind the body. A himation, wrapped about the body, covers the arms and hands, reaching much below the knees. Below it is seen the chiton, which conceals the feet.

344. Red terra cotta, exterior now gray. Height, $7\frac{1}{8}$ inches. Found at Kiti at the temple of Artemis Paralia, near Larnaca.

Figure of a pregnant woman; turban-like crown about the forehead, extending behind and above the ears; hair gathered in a knot behind. Himation wrapped closely twice about the neck and body, covering the arms; below it is seen the chiton, falling to the feet and concealing them. The right arm rests on the hip; the left, bent forward, supports the falling folds of the himation. Traces of blue and red color on the clothing. Broken, and reset; a large piece on the right side, near the waist, broken away.

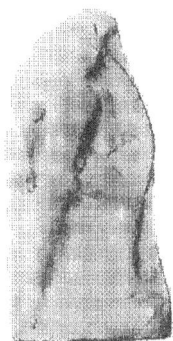

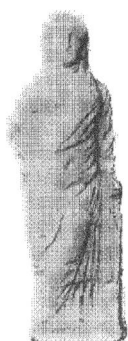

™L

Plate XLIV.

---

345. Red terra cotta, exterior now salmon-colored, with white spots. Height, 4⅛ inches. Found at the temple of Apollo Hylates, near Curium.

Beardless figure, seated on the ground, probably a devotee at a temple or shrine. Features well moulded. Head with crown of leaves that are rather heavily moulded. Right leg on the ground, bent under the body; the left bent so as to rest on the foot. Left hand on the knee; bracelet on the wrist. Most of the right arm broken away, but it probably had the hand resting on the ground. Upper part of the body clad in a very short, close-fitting, short-sleeved tunic, reaching scarcely to the waist. The rest of the body naked. A string of amulets hangs from the left shoulder across the body and under the left arm. Head broken off and reset. Slight abrasions about the base of the figure.

346. Gray terra cotta. Height, 4¾ inches. Found at same place as last.

Beardless male figure, seated on the ground (or on a divan?), the legs widely separated, the left bent almost under the body, the right bent at the knee and resting on its foot; left hand on the ground, right hand on right knee. Head with coronal semicincture, beneath which the hair appears over the forehead in short bandeaux. Faint traces of a throat necklace. Some loose drapery covers nearly all the right leg and forearm, trails on the ground in front, then, extending about the body behind, enwraps the left hand and arm a little above the elbow. Otherwise the figure is nude. This figure is much more finely wrought than the preceding. Small traces of red and brown color here and there. Nose, ears, left knee, and right foot much abraded; right hand broken away.

347. Red terra cotta. Height, 4⅜ inches. Found at same place as last.

Naked, bald-headed male figure, in nearly the same posture as the last, but the legs not widely separated. Each hand on its corresponding knee. Base three-sided, the rear side covered. Very slight chipping and abrasion here and there.

PLATE XLIV. CONTINUED.

348. Red terra cotta. Height, 3⅛ inches. Found at same place as last.

Fragment of figure of almost exactly the same description as No. 346, but much more rudely made in every way; on a lower base; and the body leaning forward a little. Covered with red color. Head and neck broken away.

349. Red terra cotta. Height, 7⅝ inches. Found at same place as last.

Naked, corpulent male figure, in nearly the same posture as No. 346, but seated on a low divan (which forms the base), one arm of which he grasps in his left hand. Hair short and curly, What seems to be the fragment of a serpent twines about the right ankle. Right arm broken away at the shoulder; a large portion broken away from the middle of the base up to the loins in front, carrying away the left foot; by the same fracture also much of the back is broken away. Features somewhat abraded.

350. Buff terra cotta. Height, 2⅝ inches. Found at same place as last.

Youthful hermaphrodite, seated on the ground (an oval base), with the knees bent so that the left foot is beneath the body, the right a little away from it. The left hand rests easily on the ground; the right, upraised, with arm extended, holds a corner of the mantle, and seems to be shaking it in sport at a shaggy dog with curled tail, which is jumping up above the right foot of the figure. Hair in a roll over the forehead; a long wavy tress also visible behind on the left shoulder. A large peplos or mantle covers the back of the head, falling and spreading behind, forming a sort of cushion beneath the figure, as well as a *quasi* curtain falling in folds from the extended right arm. Finely wrought in good Greek style. Features and right hand somewhat abraded. A crack in the base.

351. Dark buff terra cotta. Height, 3⅓ inches. Found at the Salines near Larnaca.

Winged Eros, seated on an oval base, in nearly the same posture as No. 347, but the right hand holds a bird (duck?) on the knee. Drapery loosely wrapped about the left arm below the shoulders, covering in its folds both legs and feet, but not the right arm. Good Greek style. Hair colored reddish brown. Tip of left wing broken away; slight chippings about the base.

31

348

34

351

PLATE XLV.

— — — —

352. Red terra cotta, exterior now buff-colored. Height, 4$\frac{1}{4}$ inches. Found at Dali.

Fragment of bearded warrior; helmet with close crest curving forward, and long cheek-lappets descending upon the shoulders; the lappets with small round pieces (rosettes?) just below the body of the helmet. Left arm broken away at the shoulder; but the stump as if it had supported a shield. Right arm uplifted as if to hurl a weapon; but the hand is broken away. Body very long, appearing as if it never had legs, but was once part of a chariot group. Lower part of body broken away.

353. Buff terra cotta. Height, 6⅞ inches. Found at the temple of Apollo Hylates near Curium.

Figure with rude cylindrical body not made on a wheel; thick neck; close-fitting cap on head, with wreath of leaves in front, beneath which the hair appears in wavy puffs or rolls over the forehead. Face Greek, finely moulded. Ear-rings, half a turn of wire with flat button-like ends. Traces of red color here and there. Small abrasions on nose and hair. The larger portions of both arms, and small portions at base, broken away.

354. Buff terra cotta. Height, 5¼ inches. Found at same place as last.

Figure with undulating cylindrical body; rude indication of feet, as if pushed forward under a concealing robe, hinting that the figure is draped. Face Greek, finely moulded; hair in puffy waves; on the head a high cap, its top flattened and bending over toward the left; with an ample neck-screen, spreading widely on each side, falling in front of the right shoulder nearly to the waist, but behind the left shoulder, though traces show that it once extended over the left arm to the elbow. Right arm broken away near the shoulder; part of the neck-screen broken away on the left; left arm broken away below the elbow; slight abrasions at the base.

355. Red terra cotta. Height, 4$\frac{1}{4}$ inches. Found at same place as last.

Fragment of bearded figure with cylindrical body, perhaps a warrior. Head-dress, a cap

PLATE XLV. CONTINUED.

or helmet, with point curved down forward upon the wide, nearly vertical brim. The brim surrounds the face like a halo, bearing, just below the hardly visible ears, two large disks, or buttons, below which it spreads away to hang over the shoulders like untied lappets. Hair and beard wrought with incised marks. Both hands broken away. Face, ends of hat-brim, and sundry portions of the body, somewhat chipped or abraded.

356. Red terra cotta. Height, 3¾ inches. Found near Dali.

Figure of Hercules, bearded, wearing the lion's skin; its head his helmet, and its paws about his neck. The figure wears also a tunic reaching nearly to the knees. The uplifted right arm holds in the hand an object, that may be the remains of a club, or a stone; the left arm, now broken away at the elbow, clasps some object not clearly recognizable. Figure very muscular and long-limbed, with stout neck. Lion's skin colored red, except the ears, top of head, and moulded disks of eyes, which are brown. Belt about the waist indicated in red color; eyes and beard colored brown; traces of brown and red throughout. Small pieces broken off the head of the lion's skin. Feet broken away.

357. Gray terra cotta. Height, 4⅝ inches. Found at the Salines near Larnaca.

Fragment of female figure, closely draped and muffled; all the face covered, except the forehead and eyes. Head-dress long, inclined backward, pointed; from it, a veil falls behind upon the shoulders. Beneath the drapery the right arm holds the covering to the face; the left arm hangs down, with elbow bent so that the hand grasps or holds some of the clothing from within. Surface much injured and broken, so that the detail of the drapery cannot be further made out. What at first sight seem to be folds of drapery are only deep scratches or furrowed abrasions.

358. Red terra cotta. Height, 3 inches. Found at the Salines, near Larnaca.

Fragment (head and bust) of figure of a faun, or satyr; with ample beard, long moustaches, snub nose, retreating forehead, and high bestial ears (of which the tips are broken away). On the head a low round cap, beneath which the hair protrudes over the forehead. Traces of some pendant on the left side of the cap. Stumps of arms show that the latter were held forward, but in what manner or for what purpose is not apparent. Whole figure once colored red.

PLATE XLV. CONTINUED.

359. Red terra cotta, exterior now gray. Height, 2¼ inches. Found at the Salines, near Larnaca.

Fragment of a faun, or satyr, bearded; apparently running in pursuit of his natural prey; hands upraised to head; rudimentary horns. Left hand, and all of the body below the loins, broken away. Traces of red color here and there.

360. Red terra cotta. Height, 2¼ inches. Found at the Salines, near Larnaca.

Fragment of a faun, or satyr; features more goatish than in No. 358. Horns spread in a curve over the forehead, from the root of the nose, between the eyebrows. Forehead wrinkled; hair above it; beard with curling ends; long bestial ears on the sides of the head. A head-dress present, but undefinable because of breakage. Body shaggy below the beard. On the breast is the head of some animal (panther?) for ornament. Part of right arm, left side, and all of figure below the waist, broken away.

PLATE XLVI.

361. Red terra cotta. Height, 2⅝ inches. Found at the temple of Apollo Hylates, near Curium.

Fragment of an Eros; seated on the ground, holding a round object in the right hand. Left side, from shoulder downward, broken away, as well as the lower portion, below the left foot and bent left knee. Of good Greek style.

362. Buff terra cotta. Height, 2⅛ inches. Found at same place as last.

Figure of curly-haired boy, seated on the ground; the left knee bent horizontally, and the right bent perpendicularly; almost as in No. 370, Plate XLVII. Right hand on right knee, left hand resting on the ground. Surface much defaced.

363. Red terra cotta. Height, 4⅝ inches. Found at the temple of Artemis Paralia, at Kiti, at the Salines near Larnaca.

Eros, seated on a rock or bank, holding a swan. On the head a low pilleus; about the neck, a chlamys clasped in front, hanging behind the right shoulder, covering the left shoulder and arm to the hand, and trailing off in folds to the rear. Hair beneath the pilleus, on the sides, colored red; traces of red and brown coloring on the rocky seat, the folds of the chlamys, and the wings. Wings partly broken away. Of good Greek style.

364. Red terra cotta, exterior now gray. Height, 2 inches. Found at same place as No. 361.

Upper portion of a figure of a boy, holding a bird (goose?) in his left arm. Surface much defaced, but there is still visible a himation over the head and shoulders.

365. Red terra cotta, exterior now gray. Height, 2½ inches. Found at same place as last.

Upper portion of figure of a curly-headed boy, feeding a duck, which he holds in his left arm. Head inclined a little to the right, face with a plain expression of pleasure. Clad in a close-fitting, short-sleeved chiton. Of good Greek style.

PLATE XLVI. CONTINUED.

366. Red terra cotta. Height, 2⅞ inches. Found at same place as No. 363.

Child, in half reclining posture; long curly hair; resting on a mantle which is thrown over the legs from the right hip. Head inclined to the right. Left arm, bent at elbow, rested on support (corner of couch?), but is broken off at the elbow. Right arm broken away. Broken, and reset. Of good Greek style.

367. Buff terra cotta. Height, 3₁¹₆ inches. Found at same place as No. 361.

Seated child, of description similar to the fragment No. 337, Plate XLII. ; abundant curly hair, parted in the middle; right hand on the edge of the seat; left hand holding a bird in his lap. Of good Greek style. Broken, and reset.

368. Gray terry cotta. Height, 2½ inches. Found at same place as No. 363.

Fragment of a figure of a woman holding with her left hand a nude child by the shoulder. Only the head and a portion of the left side of the body of the child are preserved, along with a fold or two of the woman's drapery.

PLATE XLVII.

369. Red terra cotta. Height, 2¾ inches. Found at the temple of Apollo Hylates, near Curium. (This place is called by the natives Apellon, formerly Hyles.)

Upper part of figure of a curly-headed boy; a himation wrapped in tight folds about the neck, and held together by the left hand from within. Of good Greek style.

370. Red terra cotta, exterior now grayish salmon color. Height, 2½ inches. Found at same place as last.

Figure of a child, seated on the ground, with right knee bent perpendicularly, holding in the right hand a ball or fruit; the left arm partially supporting the leaning body. Surface much worn. Head broken off, and reset.

371. Red terra cotta. Height, 2⅞ inches. Found at same place as last.

Fragment of relief. Female seated on a rock, or pile of rocks; hair dressed high above the head; mantle (himation) blowing away from her shoulders behind, but still over her left forearm; otherwise nude. The left hand holds what seems to be a torch; the right arm, stretched a little away from the body, rests on the rock; heavy bracelet on right wrist. Seems to be Demeter on her return from Hades.

372. Red terra cotta. Height, 3¼ inches. Found at same place as last.

Fragment of upper portion of nude boy, wearing a low tiara on the head. Shoulders raised and arms bent forward; head turned a little towards the right. Style and position almost exactly that of the Ares of the Villa Ludovisi at Rome. Fine Greek style and execution.

373. Red terra cotta, exterior now gray. Height, 2¼ inches. Found at same place as last.

Fragmentary trunk and limbs of nude and obese figure, seated on the ground; the posture much resembling that of No. 370. This class of figures much resemble a certain kind of figure in both stone and terra cotta, found in great abundance near the same spot.

PLATE XLVII. CONTINUED.

374. Gray terra cotta. Height. 2¼ inches. Found at same place as last.

Fragment. Head and left shoulder of a boy; clad in a himation which was held to the right side of his neck by the left hand, the left arm being bent across the body from the shoulder beneath the garment. Of good Greek style.

375. Red terra cotta, exterior now brownish gray. Height, 3 inches. Found at same place as last.

Female figure with straddling legs; himation wrapped over the shoulders and covering the arms, held together by the right hand raised from within, its folds hanging on the left side, but grasped by the left hand; both arms bent at the shoulders. The himation falls to the knees behind, but exposes the legs and lower part of the body in front.

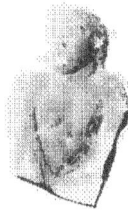

PLATE XLVIII.

376. Red terra cotta, exterior now grayish salmon color. Height, 11 inches. Found at the Salines near Larnaca, near the ruins of the temple of Artemis Paralia.

Demeter enthroned; with two attendants standing beside her. The goddess wears the ornamental calathus, which has a row of rosettes about its middle and a border of points at top. Under the calathus is a light, short peplos, that seems to descend no farther than the shoulders; beneath which is seen the hair, in rolls about the forehead, and in curls at the sides. Her face is full, and well moulded. Dress, an under tunic, or robe, reaching to the feet, fringed at the bottom; its folds appearing on and between the limbs below the knees, and thickly at the ankles. Over this tunic, a mantle, of thick, ribbed stuff, as if a series of minute parallel folds, covers the shoulders and upper part of the body, hanging down on the right-hand side, within the chair, to the foot, covering all the right arm except the hand (which rests on the right knee); and disclosing, while it covers, the right breast; covering the left arm nearly to the elbow; and is held by the left hand nearly to the chin, concealing the left breast. The feet are clad in shoes.

The throne has ornamental pieces at the sides at top, as if rounded ends of prolongation of the two uprights and the crossbar. The arms of the throne are ample; its front legs tapering downwards with an interrupting swell at the middle; appearing to have been fluted.

The attendants, one on either side of the throne, have puffed and pointed frisures; a long peplos flowing down behind and over the shoulders; wide double-collar necklaces; and a thin garment reaching to and almost covering the feet, displaying the form, and also showing its own folds. The right arm of each hangs at the side, grasping a fold of the peplos. The left arm carries a *cista* (or box of sacrificial implements or material). Surface somewhat defaced in spots and small chips here and there. Traces of red color throughout.

PLATE XLIX.

All found at the temple of Artemis Paralia, at the Salines, near Larnaca.

377.   Red terra cotta, exterior now gray salmon color.   Height, 8¾ inches.

Figure either of an attendant of Demeter, or else of her personification in a religious procession.   Dress in general the same as that of the attendants in Plate XLVIII., No. 376, except that the head has a smooth calathus over the peplos, and the hair is in rolls over the forehead and in curls at the sides.   Traces of red color throughout.   Head broken off, and reset.

378.   Red terra cotta.   Height, 8½ inches.

Figure of same description as the last, except that the calathus shows the detail of basket-work, and the hair is a little more ample.   Head broken off, and reset.

379.   Red terra cotta.   Height, 7¾ inches.

Demeter on her throne.   Head-dress a calathus, showing coarse basket-work; from its top a peplos falling behind the back, and gathered in folds across the lap, where it is held by the weight of the right hand, which rests upon it over the knee.   The other visible garment is a rather close-fitting, short-sleeved tunic, reaching nearly to the ankles.   Triple necklace.   Left hand holds a fruit to the breast.   Bracelets and armlets.   Feet with shoes.   Throne less elaborate than that of Plate XLVIII., being merely an arm-chair with ornamental shoulder-pieces ; the latter now broken away.   Traces of red coloring everywhere.   Surface somewhat chipped and defaced.

380.   Red terra cotta, exterior now gray.   Height, 5¼ inches.

Fragment of enthroned Demeter ; the throne without shoulder-pieces, and a little more than a short-armed chair.   Head wanting, but part of a peplos is seen drawn under the arms and resting in folds on the lap.   Dress, a tunic with diploïs ; the former low in the neck, without sleeves, reaching to the feet, and having long fringes on its lower edge.   Arms bare ; the left resting its hand on the knee ; the right, bent at the elbow, and holding a flower between the breasts.   Feet with shoes.   Surface somewhat abraded.

377

PLATE L.

———

Fragments of statuettes, all but No. 388 representing Demeter. She is seated on a chair or throne, of which two ornamental wing- or fan-like projections appear behind the shoulders. All were found at the Salines near Larnaca. All of red terra cotta, the exterior now gray or grayish salmon color.

381. Height, 4⅝ inches.

Demeter; woven cylindrical calathus; peplos; collar-necklace; left arm raised to neck beneath the enwrapping himation. Throne ornament broken away on right side.

382. Height, 3⅞ inches.

Demeter; like the last, except that the himation has been thrown off on the right, disclosing sleeved chiton; armlet on the right arm. Throne ornament broken away on the right side.

383. Height, 5⅜ inches.

Demeter, similar to No. 381, but calathus adorned with rosettes. Throne ornament broken away on left side.

384. Height, 2½ inches.

Demeter; in a chair without calathus; but peplos over the head, falling over the shoulders, gathered up in the hands; which are clasped together over the lap. She is apparently nursing a child, probably either Persephone or Demophoön; but the child, the arm on which it rests, and the breast, are beneath the peplos. The head bends a little forward. Surface much defaced, and corner ornaments of the chair broken away; but, notwithstanding the defects, a very graceful figure.

385. Height, 3¼ inches.

Demeter, similar to No. 382, but a fruit held in the right hand.

386. Height, 6⅜ inches.

PLATE L. CONTINUED.

Fine figure of Demeter, of same description as No. 381, except that the calathus is adorned with a wreath of rosettes and a border of ornamental points; the peplos also less ample.

387. Height, 3¼ inches.

Demeter, similar to No. 382, except that the himation is thrown off at each side. Left arm, and throne ornaments on each side, broken away.

388. Height, 1⅝ inches.

Fragments of two personifications or attendants of Demeter, with cylindrical calathi; one large peplos thrown over both, and drooping in festoon-like folds between the two.

389. Height, 3¼ inches.

Demeter, similar to No. 382; but triple necklace, triple armlet on right arm, and left hand visible as it holds the himation by its edge. Throne ornaments broken away on both sides.

PLATE LI.

All found near the temple of Artemis Paralia, at Kiti, or the Salines, near Larnaca.

390. Red terra cotta. Height, 2⅝ inches.

Fragment of figure supporting on the head, by the left arm, some object which is too much broken away to be determinable. Head with round cap (pilleus). Probably the right arm (now broken away) assisted in supporting the object. Good Greek style.

391. Red terra cotta. Height, 2⅝ inches.

Upper portion of female figure, of good Greek style ; seated on the ground. Head and body enveloped in (peplos or) himation, which covers all of the face below the nose, yet disclosing the left arm resting in the lap, and the right hand raised to the chin to hold up the edge of the mantle.

392. Gray terra cotta. Height, 2⅚ inches.

Seated female figure, clad in chiton and himation, the latter hanging over the left shoulder and across the lap. Colored red throughout. Head, forearms, and tips of feet broken away.

393. Red terra cotta, exterior now grayish salmon color. Height, 2 inches.

Head and mutilated bust of a female figure ; of good Greek style. Hair in twisted rolls over the forehead ; ear-rings ; peplos over the head, probably supported by a pointed head-dress, whose tip is now broken away. Clothing apparently a chiton.

394. Red terra cotta, exterior now gray. Height, 2½ inches.

Fragment of draped female figure, of good Greek style. Right arm raised to the head, to support some object that, with the forearm, is nearly all broken away ; but its base is round, and suggests a water jar. Hair in rolls over the forehead. Clad in a sleeveless chiton. Left arm missing.

395. Red terra cotta, exterior now gray. Height, 2¼ inches.

Mutilated fragment of female figure, of good Greek style ; hair in curls about forehead ;

PLATE LI. CONTINUED.

right arm, now broken away, was probably in same position as in No. 394; left arm hangs at the side. Dress, a sleeveless chiton.

396. Height, 2⅜ inches.

Fragment of female figure, of good Greek style; flat, low, cylindrical head-dress with peplos falling behind; hair in twisted rolls about the forehead; left arm hangs at the side, wearing an armlet; naked to the waist, below which some drapery is visible. The left hand of another figure rests on the right shoulder, indicating that this is the left-hand figure of a group.

397. Red terra cotta. Height, 2⅝ inches.

Upper portion of female figure, of good Greek style. High head-dress (broken), from which falls a peplos; dress, a thin chiton and diploïs, revealing the form as far as below the waist, where a himation encircles the body in thick folds, and is grasped by the hands at each side. Left arm broken away.

398. Red terra cotta, exterior now gray. Height, 1½ inches.

Fragment. A *cista*, or casket or box, such as was carried in processions by women personifying a goddess, or by her attendants. Similar *cistæ* are seen in the left hand of several entire statuettes (*e. g.* Pl. XLVIII., 376; XLIX., 377, 378), where they belong to the worship of Demeter.

399. Red terra cotta, exterior now brownish gray. Height, 4¼ inches.

Fragment of a group; sphinx, with female head, beside a draped sitting human figure; the latter not unlikely Demeter.

400. Red terra cotta, exterior now yellowish brown. Height, 2⅜ inches.

Fragment of representation of Demeter, with calathus and peplos; enthroned; clad in a himation. Left hand held to waist; its end broken away. Much worn and defaced.

401. Gray terra cotta. Height, 2⅛ inches.

Fragment of draped female figure, showing a triple necklace; peplos falling at the side; sleeved chiton over the right arm and breast, with clasps or buttons on the sleeve.

402. Red terra cotta. Height, 2⅜ inches.

Fragment. Head and bust of female; hair in rolls about forehead; peplos over back of head; sleeved chiton. Græco-Roman in appearance.

PLATE LI. CONTINUED.

403. Buff terra cotta. Height, 3½ inches.

Upper portion of bearded figure; apparently playing an instrument like a bagpipe, which he holds to the mouth with both hands. Pilleus on the head, over which is thrown a chlamys, which falls upon the shoulders. Abdomen naked and protuberant. Probably a Silenus. Rather rude Greek style.

404. Red terra cotta. Height, 2½ inches.

Upper portion of beardless male figure, of good Græco-Roman style; clad in tunic, with belt across from left shoulder beneath the right arm, supporting two amulets. Both arms broken away.

405. Red terra cotta. Height, 2¾ inches.

Fragment. Head and bust of female, of good Greek style; over the head a peplos falling in folds at each side of the neck, and behind the back. Simple throat-necklace of beads or pearls. Clad in chiton, with himation over the left shoulder and arm. Apparently seated in a chair; but too much is broken away to allow of certainty. Right arm, and a large portion near the left eye, broken away.

406. Red terra cotta. Height, 3½ inches.

Fragment of female figure, in the style of personifications of a goddess, with the *cista* in the left hand. Much defaced; but apparently wearing the calathus and peplos of Demeter; necklace; close-fitting chiton revealing the form as far as below the waist; himation, surrounding the lower part of the body in sinus-like folds.

407. Red terra cotta, exterior partly encrusted with gray. Height, 2⅛ inches.

Youthful head, of good Greek style; head-dress somewhat like a Phrygian cap, with wing-like appendages at the sides, reminding one of Mercury's petasus. Curls visible at its edges. Too much worn and broken for detailed description.

408. Red terra cotta. Height, 3 inches.

Fragmentary head of female figure carrying a water jar. Hair in twisted rolls about the forehead; remnants of a hand holding the jar on the right side; ear-rings. Much worn and broken.

PLATE LI. CONTINUED.

409. Slate-colored terra cotta.   Height, 3¼ inches.

Fragment of an Aphrodite ; hooded head-dress ; and peplos falling to shoulders ; otherwise nude ; arms pendent ; legs together in one piece.   Top of head on right-hand side broken away.

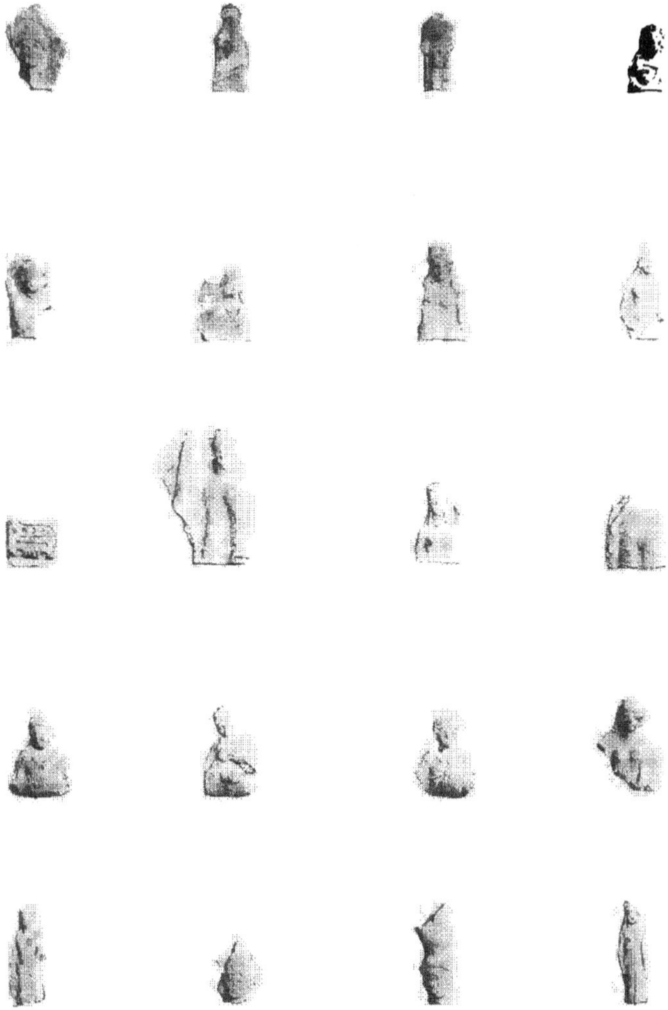

PLATE LII.

Heads from statuettes representing Demeter; or possibly some of them, *i.e.*, those with a simple woven basket quite cylindrical in shape, her attendants. As with the complete figures, Plates XLVIII., XLIX., the heads wearing the *calathus*, or basket flaring at the top, and usually adorned with rosettes and a pointed border, generally represent an enthroned Demeter. Those with the cylindrical basket (modified calathus) of woven work, are often standing figures, carrying a box (*cista*) in the left hand; a manner of representation which belonged either to maidens personifying the goddesses in religious processions, or to attendants of Demeter. All were found at the Salines, Kiti, near Larnaca.

410.   Red terra cotta.   Height, 3⅞ inches.

Head of Demeter, or woman personifying her, or else her attendant.   Basket head-dress, with peplos behind.

411.   Red terra cotta, exterior now gray.   Height, 4⅝ inches.

Head of Demeter, with high turreted calathus.   Surface much defaced.   Calathus partly broken at top.

412.   Red terra cotta.   Height, 3¼ inches.

Head of Demeter, with imperfect calathus adorned with wreaths of roses (or rosettes); hair a plain roll about the forehead, with curls at the end.   Rosette-shaped ear-rings.

413.   Slate-colored terra cotta.   Height, 3⅝ inches.

Head of Demeter, or her personification or attendant.   Basket head-dress, beneath which an ample peplos falls in wavy folds about the face.

414.   Red terra cotta, exterior now gray.   Height, 3½ inches.

Head of Demeter; high calathus.   Surface much defaced.

415.   Red terra cotta, exterior now gray.   Height, 3⅝ inches.

Head of Demeter; calathus adorned with rosettes.   Surface somewhat defaced.

PLATE LII. CONTINUED.

416. Height, 3⅛ inches.

Head of Demeter, or her personification or attendant. Basket head-dress; peplos falling behind.

417. Red terra cotta, exterior now gray. Height, 3 inches.

Head of Demeter. Calathus adorned with a border of rings below; broken away at top. Hair in twisted rolls about forehead; rosette ear-rings with long pendants. Surface defaced; nose broken away.

418. Red terra cotta, exterior now brown. Height, 3 inches.

Head of Demeter, or her personification or attendant. Basket head-dress, beneath which falls an ample peplos. Hair in close curls about the forehead. Rosette ear-rings with pendants.

419. Height, 2⅛ inches.

Head of like description with No. 410, except that the hair and ear-rings are like those of No. 412; a row of rosettes, about the calathus below, and above it a peplos.

420. Red terra cotta. Height, 3½ inches.

Head of Demeter, of nearly the same description as No. 415; but hair nearly like Nos. 412, 419. Surface very much worn and incrusted.

421. Red terra cotta, exterior now yellowish gray. Height, 2⅛ inches.

Head of Demeter, or her personification or attendant. Calathus with a row of points at top; peplos beneath it in folds, like No. 413. Somewhat worn.

PLATE LIII.

All found at the Salines near Larnaca, north of the ruins of the temple of Artemis Paralia.

422. Red terra cotta, exterior now gray. Height, 2½ inches.

Head of Demeter, or her personification or attendant; with cylindrical calathus and peplos; hair in twisted rolls about the forehead; circular ear-rings.

423. Red terra cotta. Height, 1¾ inches.

Female head, in good Greek style; hair in waves about forehead, a small puff on each side, the parting above; mantle over the head, drawn about just below the ears and under the chin. Much worn; but the mark of a thumb is still to be seen on the clay on the right-hand side.

424. Red terra cotta, exterior now gray. Height, 1½ inches.

Demeter, or her personification or attendant; similar to No. 422, but peplos in a bunch behind.

425. Red terra cotta, exterior now gray. Height, 2¼ inches.

Female head, of good Greek style; wrapped in peplos that gathers in folds over the forehead, and winds closely about the right cheek, and over the mouth and chin.

426. Red terra cotta. Height 1¾ inches.

Head of Demeter, her personification or attendant; calathus all broken away except a wreath of roses at base. Hair in rolls over forehead; ear-rings; throat necklace. Much worn.

427. Red terra cotta. Height, 1⁷⁄₁₆ inches.

Head of Demeter, her personification or attendant; low calathus; peplos.

428. Red terra cotta, exterior now grayish salmon color. Height, 1¾ inches.

Head of Demeter, her personification or attendant; ornamented calathus (broken at top); peplos visible in folds beneath it. Hair in twisted rolls about forehead. Somewhat worn; chip in the face at end of nose.

429. Red terra cotta, exterior now mottled gray and brown. Height, 1⅝ inches.

PLATE LIII. CONTINUED.

Fragment of head of Demeter, or attendant; calathus nearly all broken away; hair in twisted rolls about forehead. Much worn; tip of nose gone.

430. Red terra cotta, exterior now gray. Height, 1½ inches.

Much worn and broken head of Demeter, her personification or attendant. Remnants of calathus present.

431. Red terra cotta, exterior now grayish salmon color. Height, 1½ inches.

Head of Demeter, her personification or attendant; cylindrical calathus with ample peplos. Hair in twisted rolls about forehead.

432. Red terra cotta. Height, 2 inches.

Female head, in good Greek style; head bent a little backward and toward the right; coiffure has two large divergent bunches at top; peplos, wrapped around from behind, envelops the mouth in one turn, and the neck in another. The coiffure is like that of some cist-bearers of Demeter.

433. Red terra cotta, exterior now gray. Height, 1⅝ inches.

Head of Demeter, her personification or attendant; the calathus replaced by a flat crown of similar woven texture, upon which is a peplos. Hair in rolls, with a double puff over centre of forehead.

434. Red terra cotta, exterior now gray. Height, 1¾ inches.

Head with peplos over it; in style of attendant of Demeter without calathus. Hair in a roll, smooth over forehead, but twisted at the ends. Break across left eye and tip of nose.

435. Red terra cotta, exterior now gray. Height, 1½ inches.

Head of Demeter, her personification or attendant. A fragmentary projection indicates that the figure was either a throned Demeter, or a group of two attendants with a single peplos thrown over both, as in Plate L., No. 388.

436. Red terra cotta, exterior now gray. Height, 1½ inches.

Head, probably of a cist-bearer in a procession of Demeter. Hair plaited over the forehead, and dressed high to a point at the middle, supporting a peplos. Circular ear-rings.

PLATE LIII. CONTINUED.

437.   Red terra cotta.   Height, 2½ inches.

Female head, of good Greek style ; high puffed frisure over the forehead ; tresses falling behind the ears and down the neck.

THE GIFT

PLATE LIV.

All found at Kiti, or at the Salines near Larnaca.

438. Red terra cotta, exterior now gray. Height, $3\frac{7}{16}$ inches.

Female head, of good Greek style; hair wavy, knot behind; crescent-shaped crown.

439. Red terra cotta, exterior now dark brown. Height, $2\frac{1}{4}$ inches.

Female head, of good Greek style; wearing low crown; hair in waves about forehead, and wavy locks depending behind the ears. Much broken.

440. Red terra cotta, exterior now gray. Height, $2\frac{5}{16}$ inches.

Female head, of good Greek style; hair in a high knob over forehead; peplos falling from behind.

441. Dark gray terra cotta. Height, $3\frac{1}{16}$ inches.

Fine head of Demeter, with low calathus; peplos; long ear-drops. ·

442. Dark gray terra cotta. Height, $2\frac{1}{4}$ inches.

Female head, of good Greek style; hair in high frisure in front, and tresses falling to the neck at the sides; peplos over the hair.

443. Red terra cotta, exterior now gray. Height, $1\frac{7}{8}$ inches.

Head of female figure, perhaps an attendant of Demeter; low cylindrical head-dress, high over which an ample peplos is thrown. Hair in twisted rolls about forehead.

444. Red terra cotta. Height, $2\frac{3}{16}$ inches.

Head of Demeter. Calathus (somewhat broken) adorned with rosettes; traces of peplos.

445. Red terra cotta, exterior now gray. Height, $1\frac{7}{8}$ inches.

Head of Demeter, her personification or attendant, with cylindrical calathus and peplos underneath it. Hair in smooth rolls over the forehead; circular ear-rings.

446. Red terra cotta, exterior now gray. Height, $2\frac{1}{4}$ inches.

Head of Demeter, her personification or attendant; cylindrical calathus, over which is an

PLATE LIV. CONTINUED.

ample peplos; hair in rolls about forehead, smooth and crossing each other at top; twisted at the sides. Somewhat worn.

447. Red terra cotta, exterior now gray. Height, 1⅜ inches.

Female head, with hair in curly rolls; ample peplos. In style of Demeter or attendant. Much worn.

448. Red terra cotta. Height, 1¼ inches.

Female head, in good Greek style; hair in twisted rolls over forehead; peplos, appearing to be supported by a long hood.

449. Red terra cotta. Height, 1⅞ inches.

Head in style of Demeter, her personification or attendant; low calathus with row of rosettes; twisted rolls; peplos.

450. Red terra cotta. Height, 3⅞ inches.

Female head, of good Greek style; hair in puffed rolls with side tresses; peplos over back of head.

451. Red terra cotta. Height, 1¾ inches.

Head of Demeter, her personification or attendant; similar to No. 446.

452. Red terra cotta, exterior now gray, with traces of red coloring. Height, 2½ inches.

Female head, after the style of an attendant of Demeter; the wavy hair dressed to a point above forehead. Somewhat worn.

453. Gray terra cotta. Height, 3⅛ inches.

Female head, of good Greek style; round cap, over which falls a peplos in plaits over the forehead, and also wraps round the face from the right, enveloping the right cheek, mouth, and chin.

PLATE LV.

More or less fragmentary heads of Demeter, with calathus and peplos. Nos. 456, 457, 461–469, may be either maiden personifications of the goddess, or else representations of her attendants. All are of red terra cotta, now more or less grayish yellow, brown, or mottled on the surface. All were found at Kiti (Citium), or, more nearly, at the Salines, near Larnaca, in the neighborhood of the ruins of the temple of Artemis Paralia.

454. Height, 3½ inches. Calathus with rosettes. Hair in twisted rolls over forehead. Circular ear-rings with plummet-shaped drops.

455. Height, 3½ inches. Calathus and ear-rings as in No. 454. Hair in smooth roll over forehead; twisted at the sides.

456. Height, 3¼ inches. Calathus broken. Hair and ear-rings as in No. 455.

457. Height, 3½ inches. Calathus gone. Hair and ear-rings as in No. 455.

458. Height, 2⅞ inches. Calathus with rosettes and points. Hair and ear-rings as in No. 455.

459. Height, 3 inches. Calathus, etc., as in the last. Peplos, in this and all the preceding, over the top of the calathus, and falling behind.

460. Height, 3₁⁷₆ inches, with a single row of rosettes; peplos in folds beneath it; the rest as in No. 455. Now in the Musée du Louvre, at Paris.

461. Height, 2¾ inches. Of like description with No. 460.

462. Height, 2½ inches. Of like description with No. 454, except that the calathus is simple basket work, with points at top.

463. Height, 2⅝ inches. Of like description with No. 462, but hair in a thicker roll.

464. Height, 3¼ inches. Of like description with No. 462.

465. Height, 2½ inches. Calathus, simple basket work; hair in a thick roll over the forehead only; ear-rings circular, without drops; peplos ample.

PLATE LV. CONTINUED.

466. Height, 2¼ inches. Of like description with No. 465.

467. Height, 2 inches. Of like description with No. 454.

468. Height, 1⅞ inches. Calathus as in No. 460; peplos above it; hair and ear-rings as in No. 465.

469. Height, 2 inches. Of like description with No. 465.

PLATE LVI.

470. Height, 7½ inches. Found at Dali (Potamia).

Male head; beardless; with wreath of leaves, and hair visible in curls under the wreath. Broken away, and chipped, here and there.

471. Red terra cotta, exterior now yellowish gray. Height, 7¼ inches. Found at Dali (Potamia).

Male head; beardless; with hair in bandeaux, and above them the fragments of a wreath. Somewhat broken and chipped. Neck broken off, and reset.

472. Red terra cotta. Height, 12¾ inches. Found at the temple of Apollo Hylates, near Curium.

Rude figure; body turned on a wheel, elliptical from shoulders to neck, round and tapering somewhat from shoulders to feet. Indication of neck-tendons and collar-bone; feet, shod, project below; right arm uplifted, wanting the hand; breasts indicated by small knobs. Head thrown back; about it fragments of a wreath still remaining. Nose large, slightly retroussé. Eyes and lips strongly moulded. Colored red throughout. Left arm broken away.

473. Height, 7 inches. Found at the temple of Apollo Hylates, near Curium.

Male head, bearded; hair short and close; countenance that of a man of early middle age. Now in the Musée du Louvre, Paris.

474. Red terra cotta, exterior now yellowish gray. Height, 6⅝ inches. Found at Dali (Potamia).

Beardless male head, with prominent nose; appearance that of approaching old age. Much damaged.

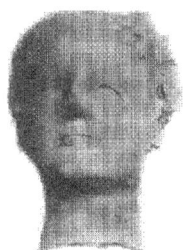

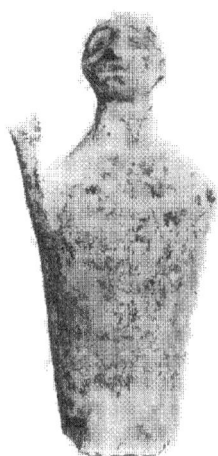

PLATE LVII.

475. Very dark-gray terra cotta. Height, 3¼ inches. Found at Dali.

Female head, in fair Greek style; face upturned. Hair in waves about forehead and down to base of neck. All except the face and front hair covered with a hood or mantle.

476. Red terra cotta, exterior now gray. Height, 3⅛ inches. Found at Dali.

Female head, of fair Greek style; coronet or diadem above the brow, and beneath it the hair in braids. Ear-rings, but their form not easily discernible. Probably a peplos descended from the head behind; but the top of the head is now broken away; the point seen in the phototype being the result of a fracture. Surface much defaced.

477. Red terra cotta, exterior now grayish yellow. Height, 5¾ inches. Found at Dali (Potamia).

Head of a young man; the thick short hair confined by a broad fillet. Well moulded. Surface much abraded; almost all the nose broken away.

478. Gray terra cotta. Height, 4¾ inches. Found at Dali (Potamia).

Youthful male head, covered by a cap with a ruffled border, presenting the appearance of a wreath; under which appears the hair above the forehead, wrought conventionally. Tip of nose and of chin broken away; and small abrasions here and there. Traces of brown color on the eyebrows, eyes, and mouth; and in a few other spots. Traces of red color throughout.

479. Red terra cotta, exterior now gray. Height, 7½ inches. Found at Dali (Potamia).

Female head with sharp features; of a frequent Cypriote type, except that the nose is aquiline. Hair in close curly bandeaux above the forehead, and covered by a peplos which hangs over the ears behind. Over this peplos is a close-fitting cap, with edge slightly upturned in a ruffle. Eyes very rudely formed. Abrasions at tip of nose, on the hair above the forehead, and in the cap-ruffle. Surface defaced in spots.

PLATE LVII. CONTINUED.

480. Red terra cotta, exterior now gray. Height, 4¼ inches. Found at the temple of Apollo Hylates, near Curium.

Full-faced, or rather, fat-faced, beardless male head ; with cap somewhat like the last, but ruffle only in front. Rude ears. Traces of brown color on features. Surface somewhat defaced. End of nose abraded.

481. Red terra cotta, exterior now grayish brown. Height, 4⅜ inches. Found at the temple of Apollo Hylates, near Curium.

Male head, beardless, full face ; nose a little retroussé ; all the features prominently wrought. Somewhat defaced and chipped.

PLATE LVIII.

482. Red terra cotta. Height, 13½ inches. Found at Soli.

Male head and neck; beardless. Features well wrought, but ears heavy, and out of shape and place. Hair in short curls, wrought by incised or stamped marks. Fragment of a crown of olive leaves about the brow. Fragments of chlamys visible on the neck, especially on the left side. Neck cracked, and reset. Surface partly covered with incrustation.

483. Buff terra cotta. Height, 9¾ inches. Found at Soli.

Beardless male head, with strong features; ears high and prominent; about the brow a wreath or crown made either of leaves or thin imitations of leaves (as in metal). Top of head left unwrought, perhaps indicating that the statue was intended for an elevated position. Much of the wreath broken away; tip of nose and of right ear abraded.

484. Red terra cotta, exterior now buff. Height, 11 inches. Found at Soli.

Head and neck of male figure; wearing a double wreath of leaves. Hair and beard wrought in incised lines. Ears rather rudely shaped. About the neck are portions of a chlamys or chiton.

485. Red terra cotta. Height, 11 inches. Found at Soli.

Head of female figure. Hair wrought in incised or stamped spirals, in ample curly masses over forehead and at the sides, covering the ears; although ear-rings are visible, in the shape of rosettes. Apparently there are remains of a peplos; but all of the head behind the hair visible on the plate, is broken away. Tip of nose, part of lips and chin, chipped away. Surface somewhat abraded throughout.

486. Red terra cotta. Height, 13½ inches. Found at Soli.

Head and neck of female figure, of same general description as the last, except that the curl-spirals are in relief. Large ear-rings, button-shaped, with a large drop. Three throat necklaces, consisting of strings of large disks, some of which are broken away. Over the head a diadem-like support for a hooded peplos, which falls down the neck behind. Top and back of head left open originally; indicating either a separately wrought piece, or else an elevated position for the statue.

—

PLATE LIX.

487. Red terra cotta, exterior now gray.   Height, 4¾ inches.   Found at the temple of Apollo Hylates, near Curium.

Head of youth; features finely moulded; hair indicated by a slightly raised portion of the material all over the head, which also appears to have been wrought somewhat in detail; but the incrustation and abrasion prevent ascertaining exactly.   The whole surface slightly marred.

488. Red terra cotta.   Height, 5⅝ inches.   Found at same place as last.

Head of youth, well moulded; nose slightly incurved, pointed; hair wrought in stamped spirals, and colored dark brown.   Eyeballs colored white, with dark brown iris and pupil; lips colored red and brown.   Other parts of the surface show traces of flesh color, beneath which is a gray.

489. Red terra cotta, exterior now yellowish gray.   Height, 4½ inches.   Found at Dali.

Male head; with curly hair, and remains of a wreath.   Somewhat worn and defaced.

490. Dark-gray terra cotta.   Height, 4½ inches.   Found at the temple of Apollo Hylates, near Curium.

Fragment of a head, comprising the face only.   Remnants of hair in bandeaux over the forehead.   Nose a little incurved, its tip broken away.   Other features full.

491. Red terra cotta.   Height, 5 inches.   Found at the Salines near Larnaca.

Front part of head of male figure, wrought in fine style.   On the head a low, flat cap, beneath which appears short curly hair.   Beard and moustaches curly.

492. Red terra cotta.   Height, 5 inches.   Found at same place as No. 490.

Fragment of head of very full-faced, youthful figure; over the head a cap or hooded mantle, with ruffled edge in front and at the sides.   Behind hangs a fragment of a veil or mantle, which may have been of one piece with the head-dress; but the breakage renders decision impossible.   Mouth open as if to speak.   Tip of nose, and portions of ruffle above forehead, broken away.

PLATE LIX. CONTINUED.

493. Red terra cotta. Height, 6¼ inches. Found at Cythrea.

Head of beardless man; prominent forehead and long, straight, pointed nose; high cheek-bones; deep wrinkles about the mouth. Hair short, curly, wrought in high relief. Ears short and defective. Exterior mottled with gray incrustation, which also covers the hair.

494. Red terra cotta. Height, 5 inches. Found at Cythrea.

Fragmentary head of a woman (?); full and regular features. Most peculiar coiffure; the hair in twisted locks, which radiate generally from a rosette of them over the centre of the fore-head; having the general appearance of a chopping sea, with its tempestuous whirlpool over the forehead. Perhaps it is a *coiffure à la Meduse*. Face originally colored with flesh tint, laid on over a layer of white, which latter now appears in spots. Ears colored red; hair, eyebrows, and eyelashes colored dark brown; eyeballs white, with dark-brown iris and pupil.

495. Red terra cotta, exterior now buff. Height, 6¼ inches. Found at Cythrea.

Male figure, beardless; low forehead, pointed nose, small mouth, chin projecting. Hair short and thick, in wavy curl bandeaux. Incrustation here and there on surface. Buff exterior probably owing to original coloring, and age.

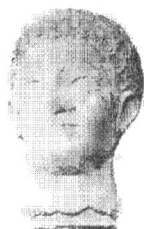

PLATE LX.

All found at the temple of Apollo Hylates, near Curium.

496. Red terra cotta. Height, 6½ inches.

Male head, beardless. Ears and chin. Apparently the head of a figure like those in Plate XXXVI. Somewhat broken away.

497. Red terra cotta. Height, 8½ inches.

Beardless male head; hair wrought by incised lines, straight and curved; crown wrought of a simple round rod, about the head.

498. Buff terra cotta. Height, 7¾ inches.

Rather youthful male head; the hair covered by a very close-fitting skull-cap or pilleus; or perhaps what seems a pilleus was only a more elevated portion, intended to be colored so as to represent the hair. Surface somewhat defaced; portions of the ears broken away.

499. Red terra cotta. Height, 8 inches.

Beardless male head; covered with a skull-cap, or pilleus, with a rounded rim, beneath which the hair appears about the forehead. This rim of the pilleus is almost precisely like the crown of No. 497, but still clearly a different affair. Rim of pilleus, and back of the head, partly broken away.

500. Red terra cotta. Height, 9 inches.

Rather youthful beardless male head, of the narrow-faced Cypriote type, with acute-angled nose and chin; hair entirely covered with a close-fitting bag, or pilleus, which seems to have a gathering-lace, like a bag, drawing up the parts over the forehead in wrinkles. On the fragmentary part of the neck and shoulder is seen the edge of a garment, probably the chiton. Nearly all the back portion, below the base of the skull, broken away.

501. Red terra cotta, exterior now buff. Height, 8 inches.

Beardless head, of Cypriote type; the head crowned with a wreath of olive (?) leaves.

PLATE LX. CONTINUED.

Surface somewhat defaced; so that it is uncertain whether the hair is enclosed in a pilleus or not.

502. Red terra cotta. Height, 7½ inches.

Rather youthful male head, beardless; with close-fitting pilleus, under which the hair appears a little over the forehead. Surface somewhat defaced.

503. Buff terra cotta. Height, 8¼ inches.

Beardless male head; thick curly hair; crown of leaves about the brows and behind the ears. Crown somewhat broken away.

504. Red terra cotta, exterior now gray. Height, 7⅝ inches.

Beardless, rather youthful, male head, with wide and high crown or wreath of leaves, in front of which the hair encompasses the forehead. Parts broken away in spots; a large portion of the back of the head thus missing.

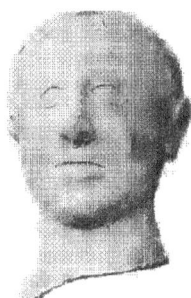

www.ingramcontent.com/pod-product-compliance
Lightning Source LLC
Chambersburg PA
CBHW031440280326
41927CB00038B/1205